ENERGY SECTOR STANDARD OF THE PEOPLE'S REPUBLIC OF CHINA

中华人民共和国能源行业标准

Specification for 3D Laser Scanning Measurement of Hydropower Projects

水电工程三维激光扫描测量规程

NB/T 35109-2018

Chief Development Department: China Renewable Energy Engineering Institute
Approval Department: National Energy Administration of the People's Republic of China
Implementation Date: July 1, 2018

China Water & Power Press

Beijing 2024

All rights reserved. No part of this publication may be reproduced, stored in a retrieval system, or transmitted in any form or by any means—electronic, mechanical, photocopying, recording or otherwise, without prior written permission of the publisher.

图书在版编目（CIP）数据

水电工程三维激光扫描测量规程 : NB/T 35109—2018 = Specification for 3D Laser Scanning Measurement of Hydropower Projects (NB/T 35109—2018) : 英文 / 国家能源局发布. -- 北京 : 中国水利水电出版社, 2024. 5. -- ISBN 978-7-5226-2597-3

Ⅰ. TV221-65

中国国家版本馆CIP数据核字第2024XT2964号

ENERGY SECTOR STANDARD
OF THE PEOPLE'S REPUBLIC OF CHINA
中华人民共和国能源行业标准

Specification for 3D Laser Scanning Measurement
of Hydropower Projects
水电工程三维激光扫描测量规程
NB/T 35109-2018
（英文版）

Issued by National Energy Administration of the People's Republic of China
国家能源局　发布
Translation organized by China Renewable Energy Engineering Institute
水电水利规划设计总院　组织翻译
Published by China Water & Power Press
中国水利水电出版社　出版发行
　　Tel: (+ 86 10) 68545888　68545874
　　sales@mwr.gov.cn
　　Account name: China Water & Power Press
　　Address: No.1, Yuyuantan Nanlu, Haidian District, Beijing 100038, China
　　http://www.waterpub.com.cn
中国水利水电出版社微机排版中心　排版
北京中献拓方科技发展有限公司　印刷
184mm×260mm　16开本　3.75印张　119千字
2024年5月第1版　2024年5月第1次印刷
Price（定价）：￥610.00

Introduction

This English version is one of China's energy sector standard series in English. Its translation was organized by China Renewable Energy Engineering Institute authorized by National Energy Administration of the People's Republic of China in compliance with relevant procedures and stipulations. This English version was issued by National Energy Administration of the People's Republic of China in Announcement [2023] No. 4 dated May 26, 2023.

This version was translated from the Chinese Standard NB/T 35109-2018, *Specification for 3D Laser Scanning Measurement of Hydropower Projects*, published by China Water & Power Press. The copyright is reserved by National Energy Administration of the People's Republic of China. In the event of any discrepancy in the implementation, the Chinese version shall prevail.

Many thanks go to the staff from the relevant standard development organizations and those who have provided generous assistance in the translation and review process.

For further improvement of the English version, any comments and suggestions are welcome and should be addressed to:

China Renewable Energy Engineering Institute
No. 2 Beixiaojie, Liupukang, Xicheng District, Beijing 100120, China
Website: www.creei.cn

Translating organizations:

POWERCHINA Chengdu Engineering Corporation Limited

Sichuan Zhongshui Chengkanyuan Surveying & Mapping Engineering Co., Ltd.

Translating staff:

ZHANG Yixi	ZHAO Cheng	ZHOU Yifei	DING Zihan
LANG Wuling	XU Haiyang	DU Xiaoxiang	LI Jianming
ZHONG Yutian	WU Zhanglei	LI Qingchun	CHEN Shangyun
HE Lei			

Review panel members:

GUO Jie	POWERCHINA Beijing Engineering Corporation Limited
QIE Chunsheng	Senior English Translator
CHEN Lei	POWERCHINA Zhongnan Engineering Corporation

	Limited
GUO Jiming	Wuhan University
LI Zhongjie	POWERCHINA Northwest Engineering Corporation Limited
QI Wen	POWERCHINA Beijing Engineering Corporation Limited
GAO Hongqi	Zhejiang Huadong Mapping and Engineering Safety Technology Co., Ltd.
YANG Xiaoyu	China Renewable Energy Engineering Institute
WANG Huiming	China Renewable Energy Engineering Institute

National Energy Administration of the People's Republic of China

翻译出版说明

本译本为国家能源局委托水电水利规划设计总院按照有关程序和规定，统一组织翻译的能源行业标准英文版系列译本之一。2023年5月26日，国家能源局以2023年第4号公告予以公布。

本译本是根据中国水利水电出版社出版的《水电工程三维激光扫描测量规程》NB/T 35109—2018翻译的，著作权归国家能源局所有。在使用过程中，如出现异议，以中文版为准。

本译本在翻译和审核过程中，本标准编制单位及编制组有关成员给予了积极协助。

为不断提高本译本的质量，欢迎使用者提出意见和建议，并反馈给水电水利规划设计总院。

地址：北京市西城区六铺炕北小街2号
邮编：100120
网址：www.creei.cn

本译本翻译单位：中国电建集团成都勘测设计研究院有限公司
四川中水成勘院测绘工程有限责任公司

本译本翻译人员：张一希　赵　程　周逸飞　丁梓涵
郎悟灵　徐海洋　杜潇翔　李建明
钟雨田　吴章雷　李青春　陈尚云
贺　磊

本译本审核成员：

郭　洁　中国电建集团北京勘测设计研究院有限公司
郄春生　英语高级翻译
陈　蕾　中国电建集团中南勘测设计研究院有限公司
郭际明　武汉大学
李仲杰　中国电建集团西北勘测设计研究院有限公司
齐　文　中国电建集团北京勘测设计研究院有限公司
高红旗　浙江华东测绘与工程安全技术有限公司
杨晓瑜　水电水利规划设计总院
王惠明　水电水利规划设计总院

国家能源局

Announcement of National Energy Administration of the People's Republic of China [2018] No. 4

According to the requirements of Document GNJKJ [2009] No. 52, "Notice on Releasing the Energy Sector Standardization Administration Regulations (*tentative*) and detailed implementation rules issued by National Energy Administration of the People's Republic of China", 168 sector standards such as *Guide for Evaluation of Vibration Condition for Wind Turbines*, including 56 energy standards (NB) and 112 electric power standards (DL), are issued by National Energy Administration of the People's Republic of China after due review and approval.

Attachment: Directory of Sector Standards

National Energy Administration of the People's Republic of China

April 3, 2018

Attachment:

Directory of Sector Standards

Serial number	Standard No.	Title	Replaced standard No.	Adopted international standard No.	Approval date	Implementation date
...						
29	NB/T 35109-2018	Specification for 3D Laser Scanning Measurement of Hydropower Projects			2018-04-03	2018-07-01
...						

Foreword

According to the requirements of Document GNKJ [2014] No. 298 issued by National Energy Administration of the People's Republic of China, "Notice on Releasing the Development and Revision Plan of the First Batch of Energy Sector Standards in 2014", and after extensive investigation and research, summarization of practical experience, and wide solicitation of opinions, the drafting group has prepared this specification.

The main technical contents of this specification include: technical preparation, control survey, airborne LiDAR scanning, terrestrial LiDAR scanning, data processing, production of digital products, and results acceptance and submission.

National Energy Administration of the People's Republic of China is in charge of the administration of this specification. China Renewable Energy Engineering Institute has proposed this specification and is responsible for its routine management. Energy Sector Standardization Technical Committee on Hydropower Investigation and Design is responsible for the explanation of specific technical contents. Comments and suggestions in the implementation of this specification should be addressed to:

China Renewable Energy Engineering Institute
No. 2 Beixiaojie, Liupukang, Xicheng District, Beijing 100120, China

Chief development organizations:

POWERCHINA Chengdu Engineering Corporation Limited

Sichuan Zhongshui Chengkanyun Surveying & Mapping Engineering Co., Ltd.

Participating development organizations:

POWERCHINA Huadong Engineering Corporation Limited

POWERCHINA Northwest Engineering Corporation Limited

POWERCHINA Beijing Engineering Corporation Limited

Chief drafting staff:

XIE Beicheng	RAO Xinggui	YAN Zhanglin	LEI Jianchao
LIU Xiaobo	LUO Yong	CHEN Shangyun	LI Xibo
LYU Baoxiong	YANG Wei	YI Juping	LI Yanling
LANG Wuling	JIA Chunhua		

Review panel members:

PENG Tubiao	FAN Yanfeng	WANG Huiming	GUO Jiming
YUE Jianping	YAN Jianguo	LIU Dongqing	SHI Jianwei
ZHANG Chengzeng	XIE Niansheng	WEN Daoping	GOU Shengguo
ZHANG Dongsheng	DENG Yong	XIAO Shengchang	GAO Hongqi
LI Shisheng			

Contents

1	**General Provisions**	1
2	**Terms**	2
3	**Basic Requirements**	3
4	**Technical Preparation**	5
4.1	Data Preparation	5
4.2	Requirements and Calibration for Instrument	5
4.3	Technical Design	7
5	**Control Survey**	9
6	**Airborne LiDAR Scanning**	11
6.1	General Requirements	11
6.2	Schematic Design	11
6.3	Data Acquisition	13
6.4	Calibration Survey	15
6.5	Data Preprocessing	16
6.6	Supplementary Survey for Scanning Gap	17
6.7	Quality Inspection	18
7	**Terrestrial LiDAR Scanning**	20
7.1	General Requirements	20
7.2	Layout and Joint Survey of Scanner Stations	20
7.3	Target Layout and Connection Survey	22
7.4	Data Acquisition	22
7.5	Data Preprocessing	24
7.6	Supplementary Survey for Scanning Gap	25
7.7	Quality Inspection	25
8	**Data Processing**	27
8.1	Airborne LiDAR Scanning Data Processing	27
8.2	Terrestrial LiDAR Scanning Data Processing	29
8.3	Quality Inspection	30
9	**Production of Digital Products**	32
9.1	Digital Elevation Model	32
9.2	Digital Orthophoto Map	32
9.3	Digital Line Graphic	33
9.4	Special Products	34
10	**Results Acceptance and Submission**	37
10.1	Result Acceptance	37
10.2	Result Submission	37

**Appendix A Record Sheet of Airborne LiDAR Data
Acquisition** 39
Appendix B Workflow of Terrestrial LiDAR Scanning 41
Appendix C Reflective Target and Observation Notebook 43
Explanation of Wording in This Specification 45
List of Quoted Standards 46

1 General Provisions

1.0.1 This specification is formulated with a view to unifying the technical requirements for the 3D laser scanning measurement of hydropower projects to ensure the quality of measurements.

1.0.2 This specification is applicable to the 3D laser scanning measurement of hydropower projects.

1.0.3 In addition to this specification, the 3D laser scanning measurement of hydropower projects shall comply with other current relevant standards of China.

2 Terms

2.0.1 3D laser scanning measurement

process that obtains the 3D coordinates, color information and reflection intensity of dense points on the target surface with a terrestrial 3D laser scanner or an airborne LiDAR

2.0.2 position and orientation system (POS)

integrated system of global navigation satellite system (GNSS) receiver and inertial measurement unit (IMU) for determining the spatial position and attitude of sensors

2.0.3 airborne LiDAR

integrated system on the air platform, composed of LiDAR, POS, digital camera and control system

2.0.4 terrestrial 3D laser scanner

integrated system on the land platform, composed of LiDAR, POS, digital camera and control system

2.0.5 scanning gap

scanned portion that has no 3D laser scanning data and exceeds a given area

2.0.6 point cloud

set of points distributed in 3D space in a discrete and irregular manner

2.0.7 density of point cloud

average number of points per unit area, usually expressed in points per square meter

2.0.8 reflective target

reference mark used for positioning and orientating during terrestrial 3D laser scanning

3 Basic Requirements

3.0.1 Prior to operation, the relevant data shall be collected and analyzed, the field reconnaissance shall be conducted and the technical design documents shall be prepared. The quality control shall be carried out during operation, and technical summary report shall be prepared after the operation.

3.0.2 The selection of coordinate system and elevation system shall comply with the current sector standard NB/T 35029, *Code for Engineering Survey of Hydropower Projects*.

3.0.3 The instruments and related equipment shall be verified, calibrated and maintained. The software used shall pass the qualification test.

3.0.4 The accuracy of 3D laser scanning measurement shall be measured by the root mean square error (RMSE), and 2 times the RMSE shall be used as the permissible error.

3.0.5 The 3D laser scanning measurement shall meet the following requirements:

1. The technical indicators of digital orthophoto map (DOM) and digital elevation model (DEM) shall be in accordance with the current sector standard NB/T 35029, *Code for Engineering Survey of Hydropower Projects*.

2. When the 3D laser point cloud data is used for feature extraction, the density of point cloud shall be in accordance with the current sector standard CH/T 8024, *Specifications for Data Acquisition of Airborne LIDAR*. When the 3D laser point cloud data is not used for feature extraction, the density of point cloud may be selected as per Table 3.0.5-1.

Table 3.0.5-1 Density of point cloud

Scale	DEM grid size (m)	Density of point cloud (point/m^2)			
		Flat land	Hilly land	Mountain	High mountain
1 : 200	0.4	2.50	5.00	7.50	10.00
1 : 500	1.0	1.00	2.00	3.00	4.00
1 : 1 000	2.0	0.50	1.00	1.50	2.00
1 : 2 000	2.5	0.25	0.50	0.75	1.00
1 : 5 000	5.0	0.10	0.15	0.30	0.40
1 : 10 000	5.0	0.10	0.15	0.30	0.40

3 The density and accuracy of point cloud used for the underground engineering mapping, deformation monitoring, etc. shall be determined according to the needs of the project.

4 The permissible horizontal RMSE of the point cloud for the digital line graphic (DLG), DEM, and DOM shall be in accordance with Table 3.0.5-2.

Table 3.0.5-2 Permissible horizontal RMSE of point cloud (mm)

Mapping scale	Flat land, hilly land	Mountain, high mountain
1 : 200	± 0.85M	± 1.10M
1 : 500 1 : 1 000 1 : 2 000	± 0.60M	± 0.80M
1 : 5 000 1 : 10 000	± 0.50M	± 0.75M

NOTES:

1 M is the denominator of the mapping scale.

2 The permissible horizontal RMSE of the ground point in the concealed area with mapping scale 1:500 to 1:10 000 may be relaxed to 1.5 times the value specified in the table above, but shall not be greater than ± 1.0M for the mountain and high mountain.

5 The permissible vertical RMSE of the point cloud shall be in accordance with Table 3.0.5-3.

Table 3.0.5-3 Permissible vertical RMSE of point cloud (m)

DEM grid size	Accuracy class	Flat land	Hilly land	Mountain	High mountain
5.0	1	± 0.25	± 0.88	± 2.12	± 3.54
	2	± 0.34	± 1.20	± 2.89	± 4.81
	3	± 0.49	± 1.77	± 4.24	± 7.07
2.5 2.0 1.0	1	± 0.12	± 0.35	± 1.00	± 1.41
	2	± 0.20	± 0.48	± 1.34	± 1.91
	3	± 0.25	± 0.70	± 1.95	± 2.83

3.0.6 When the terrain of key or sensitive areas changes, the rescanning shall be conducted timely to update the data.

4 Technical Preparation

4.1 Data Preparation

4.1.1 Before the technical design of 3D laser scanning, the following data on the operation area shall be collected:

1. Meteorology, communications, traffic, culture and physical geography.

2. Existing horizontal control points, vertical control points, photo control points, and quasi-geoid refinement, continuously operating reference station (CORS), etc.

3. Terrain type, ground cover type, and vegetation cover density.

4. DEM, orthophoto map, and topographic map.

4.1.2 Before technical design of the airborne LiDAR scanning, in addition to the data specified in Article 4.1.1 of this specification, relevant information on air activities and air traffic control at the available airports and round-trip routes near the operation area shall be collected.

4.2 Requirements and Calibration for Instrument

4.2.1 The type selection of 3D laser scanner shall meet the following requirements:

1. The measurement range shall meet the actual needs of the scanning work.

2. The accuracy of the laser point cloud shall meet the accuracy requirements of the corresponding mapping scale.

4.2.2 The parameter setting and calibration of airborne LiDAR scanner shall meet the following requirements:

1. The relevant parameters such as echo times, scan angle, and scan frequency shall be determined according to the terrain conditions of the operation area and the requirements for point cloud density and data accuracy.

2. The laser distance measurement accuracy and scan angle measurement accuracy shall be calibrated.

3. The zero position shall be calibrated.

4.2.3 The airborne LiDAR POS shall meet the following requirements:

1. An aviation GNSS receiver with high dynamic, high-precision dual-frequency or multi-frequency data receiving capability shall be used,

and the sampling frequency is not lower than 1 Hz.

2   For non-low-altitude flight data acquisition, the permissible RMSE of IMU's roll angle and pitch angle measurements are ±0.005°, and that of yaw angle measurement is ±0.02°.

3   For data acquisition from a low-altitude flight platform, the measurement accuracy of roll, pitch and yaw of the IMU is determined based on the accuracy of the laser point cloud.

4   The IMU recording frequency shall not be lower than 64 Hz.

5   The calibration of IMU shall comply with the relevant regulations of the manufacturer.

6   The shutter opening pulse of the digital camera shall be written into the GNSS data flow synchronously and accurately, and the pulse delay should not be greater than 1 ms.

4.2.4   After the airborne LiDAR system is installed, the 3D eccentricity components of the IMU and GNSS receiver shall be measured, and the permissible error is ±1 cm. The format of the eccentricity component measurement record sheet for airborne LiDAR system installation should be in accordance with Appendix A of this specification.

4.2.5   Measuring digital cameras and their verification shall comply with the current sector standard CH/T 8021, *Verification Regulation of Digital Aerial Photographic Camera*, and non-measuring digital cameras and their calibration shall comply with the current sector standard CH/Z 3005, *Specifications for Low-Altitude Digital Aerial Photography*.

4.2.6   The calibration of airborne LiDAR system shall meet the following requirements:

1   The laser scanner, POS, and digital camera of airborne LiDAR system shall be in normal working conditions.

2   The calibration shall be performed after equipment repair, severe impacts during operation, or when the data is significantly abnormal.

4.2.7   Ground GNSS receiver used for airborne LiDAR scanning shall meet the following requirements:

1   It shall match the airborne GNSS receiver.

2   The GNSS receiving antenna shall be provided with a suppression plate or ring against multipath effect.

4.2.8   The inspection of the terrestrial 3D laser scanner shall include visual inspection, power-on check and testing.

4.2.9 The visual inspection shall meet the following requirements:

1 Components of the scanner and their auxiliaries shall be complete and intact.

2 The fastening parts are not loose.

3 The bubble level, laser or optical centralizer shall be in good condition.

4 The optical telescope shall be in good condition.

5 Other auxiliary equipment shall be in good condition.

4.2.10 The power-on check shall meet the following requirements:

1 All indicators work normally.

2 The buttons and display system work normally.

3 The digital camera works normally.

4 The power-on self-check routine runs smoothly.

4.2.11 The testing shall meet the following requirements:

1 The data transmission and recording equipment and software are complete, and the data transmission performance is good.

2 The built-in electronic compass works normally.

4.2.12 The verification and calibration of terrestrial 3D laser scanner auxiliaries shall meet the following requirements:

1 The psychrometer and barometer shall be sent to the metrological department for verification, and be used within the validity period.

2 The position relation between the digital camera and the host shall be calibrated regularly, and the parameters shall be updated.

3 The reflective targets shall be inspected regularly, and the deformed ones shall be corrected or replaced.

4.3 Technical Design

4.3.1 The technical design documents shall be prepared based on the project requirements and collected data.

4.3.2 The technical design documents shall include:

1 The background, work scope, work content, workload and required time to completion.

2   The physical geography of the operation area such as terrain, climate, communications, traffic, and for airborne LiDAR scanning, the description of the surrounding airports and airspace.

3   The quantity, form, technical indicators and utilization value of the existing data.

4   Quoted standards or technical documents.

5   Types and forms of results, coordinate system, vertical datum, scale, projection method, subdivision and numbering, basic data content, data format, data accuracy, and other indicators.

6   The type, quantity and accuracy of instruments and equipment required for the work, as well as the number and functions of data processing software.

7   Technical scheme for 3D laser scanning.

8   The main requirements for quality control and product quality inspection during operation.

9   Technical requirements for data security and backup.

10 Submission and filing requirements.

11 Attached drawings, tables and other related content.

12 OHSE requirements.

4.3.3 The technical scheme for airborne LiDAR scanning shall include the working methods, technical indicators and requirements for flight plan and operation, ground GNSS base station design, and data acquisition and processing.

4.3.4 The technical scheme for terrestrial 3D laser scanning shall include the working methods, technical indicators and requirements for the layouts of the ground scanner station, the reflective target, and the control and check points, and for data acquisition and processing.

5 Control Survey

5.0.1 The horizontal and vertical control survey shall be classified into first order, second order, third order, forth order, and fifth order, where the second order, third order, forth order, and fifth order control survey shall comply with the current sector standard NB/T 35029, *Code for Engineering Survey of Hydropower Projects*.

5.0.2 When the average side length of the horizontal control network is greater than 13 km, a first order GNSS control network should be established.

5.0.3 The vertical control survey shall take the national benchmark as the starting point, and the branch level line method may be used to measure the starting data of the vertical control. The branch level line connection survey shall meet the following requirements:

1. When the length of the branch level line is greater than 20 km, the survey shall be carried out with the accuracy of no lower than the third order leveling.

2. When the length of the branch level line is not greater than 20 km, the forth order leveling may be used.

3. The branch level line shall be double-run.

5.0.4 The indicators for first order GNSS control network shall be in accordance with Table 5.0.4.

Table 5.0.4 Indicators for the first order GNSS control network

Order	Average length between adjacent points (km)	GNSS receiver	Permissible RMSE of the length between the weakest adjacent points
First	13 - 50	± (5 mm + 1 ppm)	1/500 000

5.0.5 The decimal places for calculation of the first order GNSS control network shall be in accordance with Table 5.0.5.

Table 5.0.5 Decimal places for calculation of the first order GNSS control network

Order	Geodetic latitude and longitude (")	Side length observed values and corrections (m)	Side length and coordinates (m)	Azimuth (")
First	0.000001	0.0001	0.0001	0.001

5.0.6 The first order GNSS survey shall meet the following requirements:

1. The number of the first order points of the airborne LiDAR scanning base station shall not be less than 2, with a spacing of 13 km to 50 km. The point selection shall be in accordance with the current national standard GB/T 18314, *Specifications for Global Positioning System (GPS) Surveys*.

2. For the first order network survey, the cut-off angle of satellite elevation shall not be less than 15°, the number of observation periods shall not be less than 2, the continuous observation in each period shall not be less than 6 h with an interval of 30 s, the number of satellites available in the same observation period shall not be less than 4, the total number of satellites available shall not be less than 9, and the effective observation time of any satellite shall not be less than 15 min.

3. The first order network shall use the static method to conduct connection survey to the nearest national high-order GNSS control points, or to the international GNSS service (IGS) tracking stations or CORS tracking stations. The number of connection survey shall not be less than 3, and the RMSE of the side lengths between constraint points shall not be greater than 1/700 000.

4. The number of known control points for 3D constrained adjustment shall not be less than 3. The permissible horizontal RMSE of the base station point is ±0.1 m, the permissible vertical RMSE is ±0.1 m, and the permissible RMSE of the side length between the weakest adjacent points shall be in accordance with Table 5.0.4 of this specification.

6 Airborne LiDAR Scanning

6.1 General Requirements

6.1.1 Airborne LiDAR scanning may be used for 1 : 500 to 1 : 10 000 topographic mapping, 3D modeling, and measurement of work quantities.

6.1.2 The measurement of photo-control points and coordinate and elevation conversion points shall comply with the current sector standard NB/T 35029, *Code for Engineering Survey of Hydropower Projects*.

6.1.3 The flight permit for the scan area shall be obtained before the airborne LiDAR scanning operation.

6.1.4 For each flight sortie, the airborne LiDAR scan flight record sheet shall be filled out, and the format should be in accordance with Appendix A of this specification.

6.2 Schematic Design

6.2.1 The flight route design shall meet the following requirements:

1. The point cloud data acquisition during flight shall cover the entire survey area.

2. Scan zones shall be determined comprehensively according to the scan range and surrounding terrain relief, the safe flight distance of the aircraft used, the effective ranging distance of the laser on the key target, and the maximum straight flight time as determined by the IMU accumulative error.

3. The flight altitude shall be determined according to the airborne LiDAR technical specifications and operating conditions, the precision and density of the laser point cloud, the human visual range, and the safety altitude of the flight platform.

4. The size of the laser spot shall match the detection target. When penetrating dense vegetation, the flight altitude shall be lowered and the pulse intensity shall be increased.

5. The laser pulse frequency shall be selected according to the predetermined flight altitude and laser point elevation accuracy requirements.

6. The horizontal accuracy of the laser point cloud shall be in accordance with Table 3.0.5-2 of this specification, and the vertical accuracy shall be in accordance with Table 3.0.5-3 of this specification.

7   For airborne LiDAR with roll compensation and variable view-field angle, the lateral overlap of the route should reach 20 %, and at least 13 %. For airborne LiDAR without such functions, the lateral overlap should be increased.

8   The density of the laser spots shall comply with Article 3.0.5 of this specification.

9   The flight routes within the same zone may have different flight altitudes provided that the data meet the accuracy requirements.

10 At least one tie flight shall be designed for each survey area.

11 When digital images need to be acquired simultaneously, the route design shall also meet the corresponding requirements.

12 The route position parameters and scanning parameters shall be uploaded to the airborne LiDAR.

6.2.2 The flight operation time shall meet the following requirements:

1   Select the time window with stable weather, less clouds and low wind force, in seasons with sparse tree leaves and snow-free ground.

2   When operating in busy or sensitive airspace, get to know in advance the traffic controls and select an appropriate operating time window.

3   Within the time span of the day allowed by air traffic control, select a time window with low laser attenuation and a good GNSS geometric accuracy factor.

4   When digital images need to be acquired simultaneously, select the flight operation period suitable for image acquisition as specified by the current national standard GB/T 19294, *Specification for Technological Project of Aerial Photography*.

6.2.3 The synchronous observation by ground GNSS base stations shall meet the following requirements:

1   During non-low-altitude flight operations, there shall be at least two ground GNSS base stations for synchronous observation, and the aircraft shall maintain synchronous observation with at least one base station during data acquisition, and the effective distance between the aircraft and the ground base station shall not be greater than 30 km.

2   During low-altitude flight operations, there shall be at least one ground GNSS base station for synchronous observation.

3   CORS should be used as a GNSS base station.

4 The sampling frequency of the GNSS receiver of the ground base station shall be consistent with that of the GNSS receiver in the POS.

5 The synchronous observation data from the single ground base station may be checked by precise single point positioning (PPP) solving results. When the allowable deviation of the trajectory is ±0.3 m, the solving results of the single ground base station may be used.

6 For scanning measurements at scales no larger than 1 : 5 000, IGS precise ephemeris may be used for PPP trajectory solving.

6.3 Data Acquisition

6.3.1 The flight work shall meet the following requirements:

1 Before the aircraft takes off, get familiar with the operation range, flight requirements and alternative operation plans of this sortie.

2 Before the aircraft takes off, conduct power-on check and the POS initialization. 5 min after the aircraft lands and stops completely, check the IMU and GNSS data records and confirm that they are complete before powering off the POS.

3 30 min before the aircraft takes off, complete the checkout of the ground GNSS base station and start continuous observation, and shut down the ground station 5 min after the POS is turned off.

4 During low-altitude flight operations, the start and end time of POS and ground GNSS base station observations may be determined according to the requirements of the equipment used.

6.3.2 The calibration flight shall be carried out before the operation. The calibration flight may be conducted according to the plan recommended by the manufacturer of the airborne LiDAR. The calibration flight plan of 2 altitudes and 6 routes (Figure 6.3.3) or 3 + 3 routes (Figure 6.3.4) may be adopted as well.

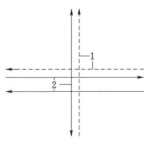

Key

1 low route   2 high route

Figure 6.3.3 Calibration flight plan of 2 altitudes and 6 routes

6.3.3 The calibration flight plan of 2 altitudes and 6 routes (Figure 6.3.3) shall meet the following requirements:

1 Two crossing routes for low altitude.

2 Two crossing routes for high altitude, one of which is a reciprocal flight with 100 % lateral overlap and one parallel with 50 % lateral overlap.

6.3.4 The calibration flight plan of 3 + 3 routes (Figure 6.3.4) shall be a route with a one-way flight with a lateral overlap greater than 50 %.

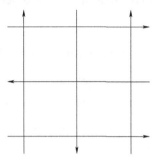

Figure 6.3.4 Calibration flight plan of 3 + 3 routes

6.3.5 The flight altitude variation shall meet the following requirements:

1 The variation of flight altitude within the same route should not be greater than 10 % of the relative flying height.

2 The deviation of the actual altitude from the design altitude should not be greater than 10 % of the relative flying height.

6.3.6 The flight speed shall meet the following requirements:

1 The flight speed shall be comprehensively determined according to the parameters such as the laser point density required for the flying direction and cross direction of a single route, the maximum instantaneous field-view angle of the system, the performance of the flight platform and the actual wind speed in flight.

2 When acquiring images simultaneously, the relevant requirements such as image overlap shall be satisfied.

3 The flight speed should keep constant during data acquisition.

4 During non-low-altitude flight scanning, the ascending and descending rates of the aircraft in the same route shall not be greater than 10 m/s.

5 During low-altitude flight scanning, the ascending and descending rates of the aircraft in the same route shall meet the requirements of the equipment used.

6.3.7 The flight attitude shall meet the following requirements:

1 The allowable deviation of roll and pitch is ±4°. As for the LiDAR with compensation function, the allowable deviation shall not exceed 2/3 of the maximum compensation value.

2 When the aircraft turns, the inclination shall not be greater than 22°.

3 The allowable deviation of the route bending is ±3 %, and the allowable deviation of the bending of the short route less than 2 km is ±5 %.

4 When flying in a straight line for more than 30 min, the aircraft shall fly in an "8" or "S" pattern before entering the survey area.

6.3.8 Airborne LiDAR scanning data acquisition shall meet the following requirements:

1 The operation shall be conducted in the time window with stable weather, less cloud and low wind force.

2 For the first route, the actual route number, altitude, direction and speed of the flight shall be verified.

3 Enter the route by alternating left and right turns.

4 When entering or exiting the route, the LiDAR and camera shall be switched on or off in time, the operation status of the automatic control equipment shall also be monitored.

5 When the data fail to meet the requirements during the acquisition, flying shall be performed again in time.

6 Tie flight shall be performed before the completion of the operation in zones.

7 All data shall be downloaded in time and backed up.

6.3.9 When the laser point cloud data and image data do not meet the requirements, the supplementary flight shall be performed.

6.4 Calibration Survey

6.4.1 The survey accuracy of the control point, laser correction point, and checkpoint of the calibration field relative to the adjacent GNSS base station shall meet the following requirements:

1 The permissible horizontal RMSE is ±0.1M mm.

2 The permissible vertical RMSE is ±1/10 of the basic contour interval.

6.4.2 The selection and survey of the calibration field shall meet the following requirements:

1 A flat and hard surface with a size of not less than 20 m × 80 m (length × width) should be chosen.

2 The calibration field shall be connection-surveyed to GNSS base stations, and the measurement error shall comply with Article 6.4.1 of this specification.

6.4.3 The selection and survey of laser correction points shall meet the following requirements:

1 A flat, hard surface such as a road, airport runway or the flat roof shall be chosen.

2 The points should be distributed over the survey area, with no less than 9 points per flight.

3 The points shall be connection-surveyed to the adjacent control points of a certain order, and the survey accuracy shall comply with Article 6.4.1 of this specification.

6.5 Data Preprocessing

6.5.1 Data preprocessing shall include the following:

1 Decode the raw data.

2 Integrate and process the POS data, ground base station observation data, and base station control point data of the same flight to solve the trajectory files.

3 Generate the point cloud data that meets the requirements, with the aid of the LiDAR scanning data.

6.5.2 POS data processing shall meet the following requirements:

1 The results of various base stations shall be in the same coordinate and elevation system.

2 The ground GNSS base station and airborne GNSS observation data shall be processed by the post-processing precise dynamic survey mode to obtain the 3D coordinates of the airborne GNSS at each moment of the flight.

3 When a CORS system tracking station or a single base station is used, the 3D coordinates of the airborne GNSS at each moment shall be checked by the results of PPP solution.

4 When the accuracy of PPP meet the requirements of the mapping scale,

the results of PPP solution may be directly used as the 3D coordinates at each moment of airborne GNSS.

5   The real-time trajectory file shall be generated by fusion of the resolution results of airborne GNSS, IMU data and eccentricity component values.

6   The quality of POS data results shall be comprehensively evaluated by indicators such as two-way solution difference, GNSS positioning accuracy and data quality factor.

7   After the processing is completed, the results of the qualified trajectory file shall be exported.

6.5.3 When using the calibration data to correct the deviation of the setting angle and laser ranging, the following requirements should be met:

1   The method and software recommended by the equipment manufacturer should be used, or the correction may be solved one-by-one manually.

2   If the software has self-calibration function, the correction may be solved when the flight strip is spliced.

6.5.4 Use the trajectory file results, LiDAR scanning data, system calibration data and system configuration files to solve and generate 3D point cloud, which should be stored in LAS or ASCII format.

6.5.5 The flight strip splicing shall meet the following requirements:

1   The horizontal RMSE of point cloud data for the same point between different flight strips shall be in accordance with Table 3.0.5-2 of this specification, and the vertical RMSE shall be in accordance with Table 3.0.5-3 of this specification.

2   When there is a systematic error with the point cloud data, laser correction points shall be used to correct the error before the flight strip splicing.

6.5.6 The color cast and exposure of the image shall be adjusted to make the color hue of the image even and true. The image exposure time shall be extracted to solve the image trajectory file.

6.6 Supplementary Survey for Scanning Gap

6.6.1 The supplementary survey for LiDAR scanning gap shall cover:

1   Area without laser point cloud data due to the obstruction of clouds and haze.

2   The gap found after data acquisition.

3   Area that lacks ground point cloud data caused by large and dense vegetation.

4   Areas of river bank lines and lake shorelines, etc. that lack echo data.

6.6.2   The supplementary survey shall overlap the cloud data of scanned points, the overlap range shall meet the edge matching, and the accuracy of the supplementary survey shall meet the requirements of the corresponding scale.

6.7   Quality Inspection

6.7.1   The quality inspection shall cover the survey area, flight quality, data quality and image quality, POS data processing and results.

6.7.2   Check whether the POS data, ground base station observation data, base station control point data, LiDAR scan raw data and image data are complete and backed up.

6.7.3   The POS data quality inspection shall include:

1   Measurement accuracy of eccentricity component.

2   GNSS observation data.

3   Time signal data.

4   IMU data.

5   POS data processing accuracy.

6.7.4   The data acquisition period of the ground GNSS base station shall coincide with the flight period, and the data acquisition shall be stable and reliable.

6.7.5   The point cloud data quality inspection shall meet the following requirements:

1   The coverage of point cloud data, the overlap of flight strips and the density of point cloud shall meet the design requirements.

2   The error range of the edge matching between the flight strips is ±1/10 of the basic contour interval.

3   The horizontal accuracy and vertical accuracy of the point cloud data shall meet the requirements of Table 3.0.5-2 and Table 3.0.5-3 of this specification.

6.7.6   The layout and survey of checkpoints shall meet the following requirements:

1 The checkpoints shall be evenly distributed in the measurement area, and there shall be checkpoints on different elevation surfaces and different reflection surfaces.

2 There shall be no less than 30 checkpoints in each scan zone.

3 The permissible horizontal RMSE of a checkpoint relative to adjacent basic control points shall be calculated as follows:

$$m = \pm 0.1M \tag{6.7.6}$$

where

m is the permissible horizontal RMSE (mm).

4 The permissible vertical RMSE is ±1/10 of the basic contour interval.

6.7.7 The image data quality shall meet the following requirements:

1 The image shall cover the full survey area without missing.

2 The image shall be exposed normally, the cloud shadow is small, the image is clear, and there is no color difference.

3 The image overlap shall meet the design requirements.

7 Terrestrial LiDAR Scanning

7.1 General Requirements

7.1.1 The terrestrial 3D laser scanning survey may be used for:

1. 1:2 000 and larger scale topographic mapping, production of DLG, DEM and DOM.

2. Deformation monitoring such as surface deformation, joints and cracks, cavern deformation, etc.

3. Surveying and mapping of building facade, 3D modeling of buildings, protection of cultural relics, reverse engineering, etc.

7.1.2 The selection of instruments and equipment shall be based on the task characteristics, work content and accuracy requirements.

7.1.3 The scanner stations for deformation monitoring should be set up on observation piers or fixed observation platforms with forced centering plate.

7.1.4 The workflow of terrestrial laser scanning shall comply with Appendix B of this specification.

7.2 Layout and Joint Survey of Scanner Stations

7.2.1 The layout plan of scanner stations shall meet the following requirements:

1. The layout of scanner stations shall consider the project purpose, accuracy requirements, range, topography of the survey area, traffic, and operation efficiency.

2. The position of scanner stations shall be determined according to the topography of the survey area, scanning distance, incident angle, etc.

3. For the scanning instruments equipped with automatic point cloud data coregistration, the scanning overlap of adjacent scanner stations shall not be less than 30 %.

4. For the scanning instruments based on coregistration by targets, the adjacent scanner stations shall share at least 3 target control points, and the overlap should reach 10 %.

7.2.2 The selection of scanner station location shall meet the following requirements:

1. The view field is wide, which is easy for setting up the instrument and operation.

2　The ground foundation is solid and stable.

3　The scanner station site should avoid the operation area of vehicles and large machinery.

4　When using GNSS to measure the coordinates of a scanner station, the location of the scanner station shall comply with the current sector standard NB/T 35116, *Specification for GNSS Survey of Hydropower Projects*.

7.2.3　Scanner stations may be set at different terrains according to the scanner station layout diagram (Figure 7.2.3).

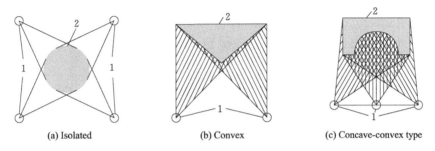

(a) Isolated　　　　(b) Convex　　　　(c) Concave-convex type

Key

1　scanner station

2　scanned object

Figure 7.2.3　Scanner station layout diagram

7.2.4　When using reflective targets for point cloud data coregistration, at least 3 scanning reflective targets shall be jointly surveyed for adjacent scanner stations.

7.2.5　The joint survey of scanner station coordinates may adopt GNSS RTK, GNSS static positioning, total station surveying, etc.

7.2.6　When the coordinates of a scanner station need to be surveyed, the point accuracy shall meet the following requirements:

1　The scanner station used for topographic surveying shall meet the requirements of Class 1 control point of topographic mapping specified in the current sector standard NB/T 35029, *Code for Engineering Survey of Hydropower Projects*.

2　For the scanner station used for surveying and mapping of building facade, 3D modeling of buildings, protection of cultural relics, reverse engineering, etc., the point accuracy shall be determined through design demonstration according to the requirements of scanning results.

7.3 Target Layout and Connection Survey

7.3.1 The making of reflective targets shall be in accordance with Appendix C of this specification.

7.3.2 The targets shall be evenly arranged in the scan area, and the number of targets shall not be less than 3.

7.3.3 When the target position is projected to the vicinity of the scanned object by the scanning distance, the accuracy of the projected point of target shall meet the accuracy of control point of topographic mapping, and each indicator shall comply with the current sector standard NB/T 35029, *Code for Engineering Survey of Hydropower Projects*.

7.3.4 Reflective targets shall be set at locations with wide field view, good line of sight, and easy to be recognized in the point cloud or image, and shall avoid strong reflection background.

7.3.5 The target diameter shall be determined considering the distance from the scanner station to the target, and it shall be 3 to 5 times larger than the expected spacing between scanning points.

7.3.6 Reflective targets shall be stable and visible during the operation. The front of the target should be perpendicular to the laser incident direction of the scanner.

7.3.7 The targets may be substituted by the points of fixed features with edges and corners that are unique and easy to identify, such as steep walls, buildings, bridges, and poles.

7.3.8 For deformation monitoring, the reference target shall be firmly set at stable place around the monitored object.

7.3.9 The coordinates and elevation survey of targets shall meet the following requirements:

1. The survey methods of the coordinates and elevation of targets shall meet the requirements of Article 7.2.5 of this specification, and the connection survey should be carried out along with the laser scanning.

2. The measurement accuracy of target shall meet the requirements of Article 7.2.6 of this specification.

3. The target measurement accuracy in deformation monitoring shall meet the requirements of deformation monitoring.

7.4 Data Acquisition

7.4.1 The spacing between scanning points shall be in accordance with

Table 3.0.5-1, and the point cloud accuracy calculated based on the maximum scanning distance shall be in accordance with Tables 3.0.5-2 and 3.0.5-3.

7.4.2 The scanner shall be used under the specified environment conditions. After startup, it shall be warmed up and keep still for 3 min to 5 min before scanning.

7.4.3 The laser incidence angle between the scanner and the scanned object should not be greater than 45°. When acquiring the point cloud data of the same target, the parameters such as scan range, point spacing, and scan frequency should be consistent.

7.4.4 The scan operation shall meet the following requirements:

1. The fine scanning of the target and the scanning of ground surface shall be completed concurrently, where there should be no power failure or scanner reset.

2. The laser head shall not be directly aimed at strong reflective objects such as prisms, mirror glass, large-area fluorescent screens at close range.

3. The scanner shall be protected from vibration during scanning operation.

4. Field scanning operation shall avoid severe weather conditions such as strong wind and haze, and should proceed when the surfaces of scanned objects are dry.

5. The observation handbook shall be filled out during field scanning operation, whose format should be in accordance with Appendix C of this specification.

7.4.5 The color image of the scanning object should be acquired together with the laser scanning, or may be acquired in good weather from the relevant scanner stations.

7.4.6 At the end of scanning operation of a scanner station, the acquired point cloud data shall be checked to ensure it is complete and correct before moving to another station.

7.4.7 Meteorological elements such as dry temperature, wet temperature, and air pressure should be collected as well during high-accuracy scanning measurements for deformation monitoring, building façade surveying, 3D modeling of buildings, protection of cultural relics, reverse engineering, etc.

7.4.8 Elements of various attributes in the scan area shall be annotated according to the following requirements:

1. The annotation takes the scanner station as unit to mark the attribute information on the corresponding area in sketch.

2. According to the surveying requirements of the corresponding mapping scale, the features, landforms and vegetation in the scan area shall be annotated.

7.5 Data Preprocessing

7.5.1 The point cloud data preprocessing shall include point cloud denoising, point cloud coregistration, coordinate transformation, etc.

7.5.2 The point cloud denoising should adopt automatic denoising by software, and manual denoising may be used.

7.5.3 When a scanner is set up at a known point and there is no reference target, the scanner shall be compensated for dual-axis tilt. When there is a reference target, the necessity of the tilt compensation may depend on the project accuracy requirements and scanner performance.

7.5.4 Either of the following two methods may be selected for coregistration of point cloud data:

1. When using target coregistration, chain coregistration and ring coregistration may be chosen. The coregistration should be carried out in the order of scanning. For coregistration, the targets with a better observation quality shall be selected, and the number shall not be less than 3.

2. For the scanner with self-leveling and orientation function, when using direct coregistration, a station may be set up at a known control point to obtain the real 3D coordinates of the scanned objects, through the orientation function.

7.5.5 When the point cloud data coregistration is completed, adjustment shall be done with the accompanying software of scanner. The permissible RMSE of point cloud data coregistration shall meet the following requirements:

1. The permissible RMSE of point cloud data coregistration for topographic surveying shall meet the requirements of Table 7.5.5.

2. The permissible RMSE of coregistration for deformation monitoring, surveying of building facades, 3D modeling of buildings, protection of cultural relics, reverse engineering, etc., shall meet the accuracy requirements of the project.

Table 7.5.5 Permissible RMSE of point cloud data coregistration for topographic surveying

Mapping scale	Permissible RMSE (cm)
1 : 200	±1.5
1 : 500	±2
1 : 1 000, 1 : 2 000	±5
1 : 5 000	±10

7.5.6 The coordinate transformation shall comply with the current sector standard NB/T 35029, *Code for Engineering Survey of Hydropower Projects*.

7.6 Supplementary Survey for Scanning Gap

7.6.1 The supplementary survey for scanning gap shall comply with Article 6.6.2 of this specification.

7.6.2 The image data should also be acquired during the supplementary survey. The image resolution and color shall be basically the same as those of the scanned images.

7.7 Quality Inspection

7.7.1 The inspection of point cloud data shall include:

1 Whether the overlap of point cloud data acquired by adjacent scanner stations meets the requirements.

2 The completeness of point cloud data and color image information.

3 The correctness of measurement results of reflective targets or feature points.

4 The correctness of the scanning data conversion.

5 The completeness and reliability of image subsidiary inspection data.

7.7.2 Rescanning shall be made when the point cloud data fails to meet the requirements.

7.7.3 A certain number of checkpoints shall be arranged in the scan area according to the following requirements:

1 The checkpoints should be evenly distributed.

2 The number of checkpoints in each scan area shall not be less than 20.

7.7.4 The quality of point cloud data shall be evaluated based on the results of the inspection according to the following requirements:

1. The difference between the model of point cloud data and the real size of the scanned object shall meet the accuracy requirements of the topographic map with the corresponding scale.

2. The density of ground points in point cloud data shall meet the accuracy requirements of the topographic map with the corresponding scale.

3. The point cloud data shall be complete.

4. The sampling accuracy of a single scanner station shall meet the design requirements.

8 Data Processing

8.1 Airborne LiDAR Scanning Data Processing

8.1.1 The airborne LiDAR scanning data shall be checked before processing, including:

1. Point cloud data.

2. Trajectory file.

3. Horizontal and vertical checkpoints.

4. Conversion between engineering coordinates and point cloud coordinates and between engineering elevation and point cloud elevation.

5. Other relevant data.

8.1.2 The point cloud data from the airborne LiDAR scanning shall be converted into an engineering coordinate system, and the conversion parameter determination and accuracy shall meet the following requirements:

1. The plane coordinate conversion may use the Bursa seven-parameter method or Helmert four-parameter method. The difference between the converted value of the checkpoint and the measured value shall meet the following requirements:

$$\delta \leq \pm 0.2M \qquad (8.1.2\text{-}1)$$

where

δ is the difference between checkpoint conversion value and measured value (mm).

2. Elevation conversion may use plane fitting method or polynomial surface fitting method. The difference between the converted value of the checkpoint and the measured value shall not be greater than 1/5 of the basic contour interval.

3. If available, the qualified refined result of quasi-geoid of the survey area may be collected and used directly.

8.1.3 Laser points shall be classified as ground points, non-ground points, and thematic points, and placed in the specified layer. The definition of point class shall be in accordance with Table 8.1.3.

8.1.4 For projects with huge storage capacity of point cloud data, the point cloud data should be processed in blocks in the principle of reducing the workload of edge matching and efficient processing according to the existing data processing software and hardware performance.

Table 8.1.3 Definition of point classes

Serial number	Point class	Definition
1	Ground point	Points that reflect the real terrain relief and that fall on the surface of the bare ground, including the points that fall on roads, squares, dams and other features that reflect the geomorphy
2	Non-ground point	Points that do not fall on the surface of bare ground, mainly refer to points that fall on the objects above the ground, such as points on buildings, vegetation, pipelines, transmission lines
3	Thematic point	Non-ground points expressed with the same type of features according to application requirements, including hydrography and facilities, settlements and facilities, transportation, pipelines, vegetation and other points

8.1.5 Laser point cloud data should be classified in three steps of denoising, automatic classification and manual editing, to obtain the results of ground point classification and thematic point classification. The following flow may be used for point cloud classification:

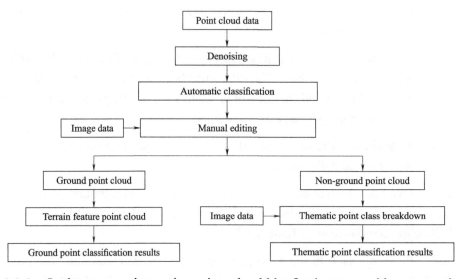

8.1.6 In data processing, noise points should be firstly removed by automatic algorithms, and then further removed manually. The noise point removal shall not affect the automatic extraction of ground points.

8.1.7 Ground points shall be extracted by classifying point cloud data using algorithms or algorithm combinations based on geomorphology, echo, and reflection intensity.

8.1.8 The non-ground points shall be classified according to the distribution characteristics of the point cloud. Features with distinct spatial distribution of point clouds, such as buildings and power lines, may be automatically extracted by software through parameter setting, and other point clouds shall be classified by manual editing.

8.1.9 Re-classification shall be done by manual editing if the automatic classification of point clouds is not correct.

8.2 Terrestrial LiDAR Scanning Data Processing

8.2.1 The data processing of the terrestrial 3D laser scanning shall mainly include color texture mapping, point cloud classification, point cloud data streamlining, and terrain feature extraction.

8.2.2 Preparation for data processing shall include:

1 Point cloud data and image data.

2 The 3D coordinates of the scanner station, target or feature control point.

3 Calibration parameters for the position relation between camera and scanner.

4 Data related to data processing and inspection of splicing results.

8.2.3 For projects that require color information, the scanned images and point clouds shall be matched and mapped.

8.2.4 Color texture mapping shall meet the following requirements:

1 When the data volume of mapping model is huge, the data shall be divided into blocks according to the image zone.

2 For images acquired asynchronously, the chromatic aberration caused by external meteorological factors shall be properly leveled in light and color.

3 The mapped block models shall be merged and spliced into a whole.

8.2.5 The point cloud data obtained by scanning shall be merged and divided into blocks according to the following requirements:

1 The point cloud data shall be merged according to the data volume of the scanner station and the terrain type.

2 The point cloud data shall be partitioned according to the data processing software and hardware performance.

8.2.6 The laser point cloud data shall be classified by type. The point cloud classification shall be done manually on the basis of automatic classification according to the following requirements:

1 The point clouds shall be classified automatically by intensity of reflection, number of echoes, RGB information, etc.

2 Unreasonably or incorrectly classified points shall be re-classified manually.

8.2.7 Point cloud data streamlining shall meet the following requirements:

1 Express the complete ground surface information and meet the needs of the production of surveying and mapping products with corresponding scales.

2 Retain the characteristic points of original features and landform completely to meet the accuracy of surveying and mapping products.

8.2.8 The elements extracted from the point cloud data shall mainly include the following:

1 Features and geomorphic elements related to DLG.

2 Characteristic elements required in the project.

8.2.9 Image data processing shall meet the following requirements:

1 When there is overexposure, underexposure, shadow, color difference between adjacent images, color adjustment shall be conducted to maintain a moderate image contrast and color consistency.

2 The image distortion caused by the angle of view or lens shall be corrected.

3 When the image is matched, the image details shall be clear and the mosaic seamless.

4 The processed image should be saved in common file formats.

5 The processed image shall truly reflect the situation on the ground.

8.3 Quality Inspection

8.3.1 The classified laser point cloud data and texture information shall be visually rendered in 3D, and visual inspection shall be carried out through classification display, elevation display, section display, etc., and in case of doubt, digital images acquired simultaneously may be used for check. The checkpoints measured in field and characteristic point data may be exported into the point cloud for accuracy statistics.

8.3.2 The quality inspection of 3D laser point cloud data processing shall check whether:

1. The interpretation of the features at the matched edges of the adjacent blocks is correct for edge matching in blocks.

2. The color information texture mapping is coincident and matched.

3. The point cloud data classification is correct.

4. The surface model established by ground point data is continuous and smooth after point cloud filtering.

5. The density of point cloud data is reasonable after streamlining.

6. The landform characteristics and thematic elements of point cloud data are correctly and completely extracted.

9 Production of Digital Products

9.1 Digital Elevation Model

9.1.1 The completeness of laser point cloud classification results shall be checked before DEM production.

9.1.2 DEM producing procedures shall meet the following requirements:

1. Process the point cloud data by triangulated irregular network (TIN) or regular rectangular grid (GRID) to get the grid model.

2. Conduct elevation correction for special areas such as water areas and roads.

3. Generate and visualize DEM.

4. Crop the DEM according to the specified size.

9.1.3 The DEM quality inspection shall cover spatial reference frame, position accuracy, logic consistency, temporal accuracy, raster quality and accessory quality.

9.1.4 DEM quality inspection may use the methods such as verification analysis, comparison analysis, and field testing.

9.2 Digital Orthophoto Map

9.2.1 The completeness of laser point cloud classification results, image data and non-image data shall be checked before DOM production.

1. Image data include aerial photograph data, ground image data, etc.

2. The non-image data include camera parameter files, trajectory files, GNSS positioning data, conventional geodetic control survey data, annotation data, DEM data, etc.

9.2.2 Differential correction shall be conducted on original single image using produced DEM or digital surface model (DSM), camera parameters and exterior orientation elements of the photo, and the image is resampled to generate a single orthophoto map.

9.2.3 DOM production shall meet the following requirements:

1. Check the color hues or the chromatic aberration between adjacent photos and adjust the color.

2. Make the image smooth within the mosaic overlap strip.

3. Calculate the cropping range of image data and generate additional

information files for image data cropping.

9.2.4 DOM should be stored in uncompressed Geo Tiff format.

9.2.5 The DOM quality inspection shall cover spatial reference frame, position accuracy, logic consistency, temporal accuracy, image quality, characterization quality and accessory quality.

9.2.6 DOM quality inspection may use the methods such as verification analysis, comparison analysis and field testing.

9.2.7 In addition to Article 9.2.1 and Articles 9.2.4 to 9.2.6 of this specification, the DOM on specific projection plane produced from terrestrial 3D laser point cloud data shall meet the following requirements:

1 Use stereo model projection or point cloud projection for image correction.

2 Ensure the completeness and correspondence of point cloud data and images for color point cloud production or texture mapping, with no obvious chromatic aberration in the overlapping area of the images.

3 The scale should not be less than 1:500, and the spatial resolution shall meet the following requirements:

$$r < 0.1M \tag{9.2.7}$$

where

r is the spatial resolution (mm).

4 Image output shall consider mapping scale, and print resolution shall not be less than 300 dpi.

9.3 Digital Line Graphic

9.3.1 The completeness of laser point cloud classification results, images and other relevant data shall be checked before DLG production.

9.3.2 DLG terrain classification, selection of basic contour interval, permissible horizontal RMSE of feature point, and permissible vertical RMSE of contours and labelling points shall comply with the current sector standard NB/T 35029, *Code for Engineering Survey of Hydropower Projects*.

9.3.3 Elevation labelling points shall be placed at obvious feature point and terrain characteristic point, whose density shall be determined depending on the map load; 5 to 15 points should be labelled within each 100 cm^2 on the map. When the contour interval is less than or equal to 0.5 m, the elevation points shall be labelled accurate to 0.01 m while the rest to 0.1 m.

9.3.4 The contour shall be smoothed within the allowable accuracy range when it is generated by using the results of laser point cloud classification.

9.3.5 Extracted feature and geomorphic elements in DLG shall include:

1 Survey control points.

2 Hydrographic feature.

3 Settlements and facility.

4 Individual feature.

5 Traffic.

6 Utility lines.

7 Boundary.

8 Land category boundary, vegetation and fence.

9 Geological exploration points, hydrology, meteorology, tide station and wind mast related to hydropower projects, etc.

9.3.6 Feature and geomorphic elements should be extracted by DOM and laser scanning point cloud data or their combination, and may extracted by integrated 3D mapping software. When extracting feature elements, the point cloud boundary shall be clear and easy to identify, and the permissible error is calculated as follows:

$$\varDelta = \pm\ 0.1M \qquad (9.3.6)$$

where

$\varDelta$ is the permissible error (mm).

9.3.7 The edge matching shall not change the actual shapes and relevant positions of various features, without causing landform distortion.

9.3.8 The DLG quality inspection shall include spatial reference frame, position accuracy, attribute accuracy, integrity, logic consistency, temporal accuracy, characterization quality and accessory quality.

9.3.9 DLG quality inspection may use the methods such as verification analysis, comparison analysis, field testing and field map matching.

9.4 Special Products

9.4.1 Special products such as digital 3D model, DSM and vertical DOM may be produced by using point cloud data and image data from laser scanning.

9.4.2 The field scan of the digital 3D model shall cover the surface of the target object, completing the overall modelling of the target object by splicing the point cloud data from several stations.

9.4.3 When overlaying texture on DSM, the data acquisition shall describe image data source and time phase in metadata in addition to complying with Article 9.4.2 of this specification.

9.4.4 When overlaying texture on DEM, the data acquisition shall comply with Article 9.4.2 and Article 9.4.3 of this specification.

9.4.5 The production of plans, elevations and sections shall meet the following requirements:

1 Use the corresponding projection plane for point cloud projection according to the types of products.

2 Point cloud may be partitioned according to data size, hardware and software performance, accuracy requirements and object type.

3 Vectorization is performed based on the point cloud feature after projection.

4 When the structure dimensions cannot be obtained completely due to partial missing of point cloud, the dimensions of concealed parts may be calculated according to that of exposed parts, and the calculation shall be described accordingly.

5 The structural dimensions shall be checked on site and the relative error shall not be greater than 1/200.

6 DOM scale should not be smaller than product scale.

9.4.6 The horizontal displacement, vertical displacement, deflection and deformation during monitoring measurement shall be obtained through comprehensive induction and analysis of the extracted characteristic points, similar series points and monitoring target points as well as simultaneous image, structural section line, model and point cloud data.

9.4.7 Measurement of work quantities shall meet the following requirements:

1 The data used for measurement shall be measured on a spatial model or be overlaid and extracted.

2 Area and volume calculations may be performed using triangular mesh or grid methods.

3 The point cloud used for measurement shall fully reflect the characteristics of the object and the point cloud grid size shall be in accordance with Table 9.4.7.

Table 9.4.7 Point cloud grid size

Grid size (m)	Scope of application
≤ 0.5	Important works, storages, tunnel mining, etc.
0.5 - 1.0	Open-pit mining, stockpiling, soil piling, building elements, etc.
1.0 - 5.0	Large area, simple landform, excavation and filling earthwork of pits, ponds and reservoirs

9.4.8 The special product quality inspection shall cover horizontal accuracy, vertical accuracy, and image correctness and rationality.

9.4.9 The special product quality inspection may use the methods such as verification analysis, comparison analysis, field testing and field map matching.

10 Results Acceptance and Submission

10.1 Result Acceptance

10.1.1 The 3D laser scanning measurement project shall be subjected to "two-level inspection, one-level acceptance". The inspection and acceptance, and quality assessment shall be carried out according to the TOR or contract, technical design documents, entrusted inspection acceptance documents, sector standards and national standards.

10.1.2 The inspection, acceptance and quality assessment of mapping results shall comply with the current sector standard NB/T 35029, *Code for Engineering Survey of Hydropower Projects*.

10.1.3 The inspection report shall be prepared after final inspection and the acceptance report shall be prepared after acceptance.

10.2 Result Submission

10.2.1 Data sorting shall be carried out after 3D laser scanning measurement, including:

1. Raw data of laser point cloud.
2. Raw image data and related technical data.
3. Results and description of control points.
4. Coordinate system transformation model data.
5. Raw observation data and calculation data of control points of each order.
6. Annotated photograph data.
7. Result of field supplementary survey.
8. Orientation modeling data and inspection record book.
9. Data acquisition, data editing and inspection record books.
10. Technical design documents, technical summary report and inspection report.
11. Project contract, airspace flight permit, etc.

10.2.2 Each category of data shall be registered one by one, and a list shall be prepared.

10.2.3 The content of data file and its format shall comply with the requirements of technical design documents or the current national standard

GB/T 17798, *Geospatial Data Transfer Format*, and the storage medium shall be CD or HD.

10.2.4 Product name, production unit, production time, etc. shall be marked on the outer packaging of the medium storing the 3D laser scanning measurement results.

10.2.5 After the 3D laser scanning measurement project is completed, submittals shall include:

1 Technical design documents.

2 Digital mapping results.

3 Technical summary report.

4 Inspection report.

5 Other data required by technical design documents.

Appendix A   Record Sheet of Airborne LiDAR Data Acquisition

A.0.1   The format of eccentricity component measurement record sheet for airborne LiDAR system installation should be in accordance with Table A.0.1.

Table A.0.1   Eccentricity component measurement record sheet for airborne LiDAR system installation

Basic information	Code of survey area		Name of survey area	
	Aircraft model		Aircraft number	
	LiDAR model		LiDAR number	
	IMU model		IMU number	
	GNSS receiver model		GNSS antenna model	
	Camera model		Camera number	
	Surveyor		Tester	
	Survey date		Test date	
Component direction \ Eccentricity component	GNSS		IMU	Camera
Flight direction u (mm)				
Lateral v (mm)				
Vertical w (mm)				
Remarks				

A.0.2 The format of airborne LiDAR scanning flight record sheet should be in accordance with Table A.0.2.

Table A.0.2 Airborne LiDAR scanning flight record sheet

Name of survey area				
Participants		Recorded by		
Flight date		Weather		
Scan parameters		Camera parameters		
Start time		Stop time		
Take-off time		Landing time		
Flight route number	Start and stop number of images	Flight speed	Flight altitude	Remarks

Appendix B Workflow of Terrestrial LiDAR Scanning

B.0.1 The terrestrial 3D laser scanning should follow the workflow below:

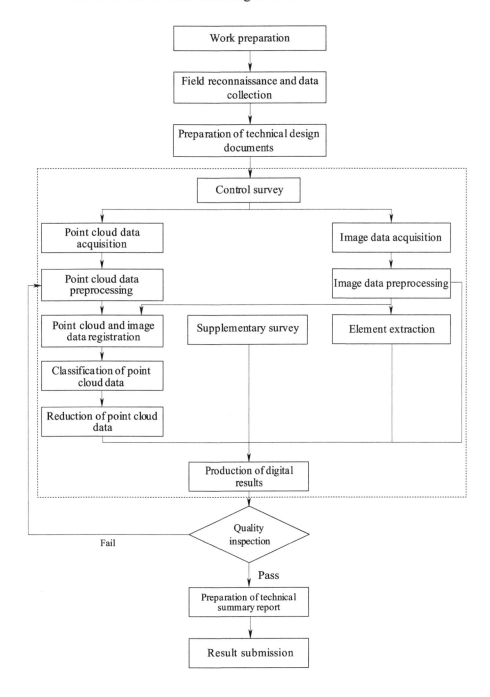

B.0.2 The field data acquisition of terrestrial 3D laser scanning should follow the workflow below:

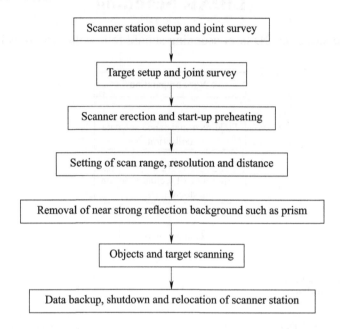

Appendix C Reflective Target and Observation Notebook

C.1 Reflective Target

C.1.1 Reflective target should be spherical and circular. The pattern of reflective target may be selected as shown in Figure C.1.1.

  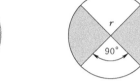 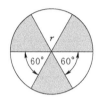

(a) Spherical target sign (b) Round X-shaped plane target sign (c) Round diamond plane target sign

Key

r radius of a spherical or circular target

Figure C.1.1 Reflective target pattern

C.1.2 The size of the reflective target may be determined according to the characteristics of the project, accuracy requirements, distance between scan points, range of the scanner, and background.

C.1.3 Target should be made of materials with strong reflection, non-fading, smooth surface and bright color.

C.1.4 The color of the target shall be greatly different from that of the object being scanned. White, black, red, blue or alternate two-color symbols may be selected according to the site conditions.

C.2 Observation Notebook

C.2.1 For the content and format of terrestrial 3D laser scanning observation notebook, refer to Table C.2.1.

C.2.2 In the notebook, the Name should be the project name. When a station is set up on an unknown point, the station name and instrument height need not be filled in. Target name and orientation point may be filled in depending on the scan mode.

Table C.2.1　Terrestrial 3D laser scanning observation notebook

Name			
Observation date		Weather	
Instrument		Instrument model and number	
Station name		Instrument height	
Target name		Target type	
Orientation point		Humidity	
Temperature		Air pressure	
Adjacent stations		Position relation of adjacent stations diagram:	
Scanning operation: (for parameters, rough scan, fine scan, scan area, etc.)		Image acquisition: (degree of overlap, quantity and others)	
Station position description:		Target position description:	
Station and target position diagram:			
Remarks:			
Surveyed by:	Operated by:	Recorded by:	

Explanation of Wording in This Specification

1   Words used for different degrees of strictness are explained as follows in order to mark the differences in executing the requirements in this specification.

    1) Words denoting a very strict or mandatory requirement:

       "Must" is used for affirmation; "must not" for negation.

    2) Words denoting a strict requirement under normal conditions:

       "Shall" is used for affirmation; "shall not" for negation.

    3) Words denoting a permission of a slight choice or an indication of the most suitable choice when conditions permit:

       "Should" is used for affirmation; "should not" for negation.

    4) "May" is used to express the option available, sometimes with the conditional permit.

2   "Shall meet the requirements of…" or "shall comply with…" is used in this specification to indicate that it is necessary to comply with the requirements stipulated in other relative standards and codes.

List of Quoted Standards

GB/T 17798, *Geospatial Data Transfer Format*

GB/T 18314, *Specifications for Global Positioning System (GPS) Surveys*

GB/T 19294, *Specification for Technological Project of Aerial Photography*

NB/T 35029, *Code for Engineering Survey of Hydropower Projects*

NB/T 35116, *Specification for GNSS Survey of Hydropower Projects*

CH/T 8021, *Verification Regulation of Digital Aerial Photographic Camera*

CH/T 8024, *Specifications for Data Acquisition of Airborne LIDAR*

CH/Z 3005, *Specifications for Low-Altitude Digital Aerial Photography*